The Last
Unknowns

ALSO BY JOHN BROCKMAN

As Author

By the Late John Brockman
37
Afterwords
The Third Culture
Digerati

As Editor

About Bateson
Speculations
Doing Science
Ways of Knowing
Creativity
The Greatest Inventions
  of the Past 2,000 Years
The Next Fifty Years
The New Humanists
Curious Minds
What We Believe but
  Cannot Prove
My Einstein
Intelligent Thought
What Is Your
  Dangerous Idea?
What Are You
  Optimistic About?
What Have You Changed
  Your Mind About?
This Will Change Everything

Is the Internet Changing
  the Way You Think?
Culture
The Mind
This Will Make You Smarter
This Explains Everything
Thinking
What Should We Be
  Worried About?
The Universe
This Idea Must Die
What to Think About
  Machines That Think
Life
Know This
This Idea Is Brilliant
Science at the Edge

As Coeditor

How Things Are
  (with Katinka Matson)

Deep, Elegant, Profound
UNANSWERED QUESTIONS
About the Universe, the Mind,
the Future of Civilization, and
the Meaning of Life

The Last
Unknowns

Edited by

JOHN

BROCKMAN

WILLIAM MORROW
An Imprint of HarperCollins*Publishers*

HarperCollins books may be purchased for educational, business, or sales promotional use. For information, please email the Special Markets Department at SPsales@harpercollins.com.

FIRST EDITION

Designed by Diahann Sturge

Library of Congress Cataloging-in-Publication Data has been applied for.

ISBN 978-0-06-289794-7

19 20 21 22 23 LSC 10 9 8 7 6 5 4 3 2 1

In memory of

Jonas Mekas

Poet, filmmaker, intellectual

1922–2019

Contents

Foreword: On Edge

It seems like yesterday, but *Edge* has been up and running for twenty-two years. Twenty-two years in which it has channeled a fast-flowing river of ideas from the academic world to the intellectually curious public. The range of topics runs from the cosmos to the mind, and every piece allows the reader at least a glimpse and often a serious look at the intellectual world of a thought leader in a dynamic field of science. Presenting challenging thoughts and facts in jargon-free language has also globalized the trade of ideas across scientific disciplines. *Edge* is a site where anyone can learn, and no one can be bored.

The statistics are awesome: the *Edge* conversation is a "manuscript" of close to 10 million words, with nearly 1,000 contributors whose work and ideas are presented in more than 350 hours of video, 750 transcribed conversations, and thousands of brief essays. And these activities have resulted in the publication of 19 printed volumes of short essays and lectures in English and in foreign language editions throughout the world.

The public response has been equally impressive: *Edge*'s influence is evident in its Google Page Rank of 8, the same as *The Atlantic, The Economist, The New Yorker,* and the *Washington Post*; in the enthusiastic reviews in major general-interest outlets; and in the more than 700,000 books sold. Of course, none of this would have been possible without the increasingly eager participation of scientists in the *Edge* enterprise. And a surprise: brilliant scientists can also write brilliantly! Answering the *Edge* question evidently became part of the annual schedule of many major figures in diverse fields of research, and the steadily growing number of responses is another measure of the growing influence of the *Edge* phenomenon. Is now the right time to stop? Many readers and writers will miss further installments of the annual *Edge* question—they should be on the lookout for the next form in which the *Edge* spirit will manifest itself.

What is the secret of *Edge*'s success? To begin with, the charisma of John Brockman, its founder and leader. Add to that his eclectic but discerning taste in the choice of participants. The two major formats of *Edge* activities are no less important. The interviews are edited to make the interviewer invisible. Masking the questioner is not new, but remarkable skill is required to elicit both clarity and depth in seamless expositions of the participants' ideas. *Edge* interviews read and sound like coherently constructed informal lectures—a surprising feat when the flow of the content is entirely driven by the interviewer's questions.

The short-essay format of the *Edge* Annual Question is a daring innovation and a striking success. Apparently, 600–1,000 words is the sweet spot for introducing one big idea. The brevity disciplines the author and allows the reader to grasp the essential point—and to remain hungry for more even as she moves to another essay.

The unifying message in the story of *Edge* is that ideas matter, and they matter to many. They can be told with elegance, sometimes with wit, never with condescension. There is a large audience eager to learn what scientists in various disciplines are up to, and a large group of scientist-teachers eager to tell their stories. And certainly, there will be more stories.

—Daniel Kahneman
New York City

Acknowledgments

I wish to thank my editor, Peter Hubbard of William Morrow/HarperCollins, and my agent, Max Brockman, for their continued encouragement. A special thanks to Nina Stegeman for her thoughtful attention to the manuscript.

Introduction
Interrogate Reality

After twenty years, I've run out of questions. So, for the finale to a noteworthy Edge *project, can you ask "The Last Question"?*

Did I say "twenty years"? My strange obsession with the idea of "Question" goes back to 1968, when I first wrote about the idea of interrogating reality[1]:

> The final elegance: assuming, asking the question. No answers. No explanations. Why do you demand explanations? If they are given, you will once more be facing a terminus. They cannot get you any further than you are at present.[2] The solution: not an explanation: a description and knowing how to consider it.
>
> Everything has been explained. There is nothing left to consider. The explanation can no longer be treated as a definition. The question: a description. The answer: not explanation, but a description and

knowing how to consider it. Asking or telling: there isn't any difference.

No explanation, no solution, but consideration of the question. Every proposition proposing a fact must in its complete analysis propose the general character of the universe required for the fact.[3]

Our kind of innovation consists not in the answers, but in the true novelty of the questions themselves; in the statement of problems, not in their solutions.[4] What is important is not to illustrate a truth—or even an interrogation—known in advance, but to bring to the world certain interrogations . . . not yet known as such to themselves.[5]

A total synthesis of all human knowledge will not result in huge libraries filled with books, in fantastic amounts of data stored on servers. There's no value any more in amount, in quantity, in explanation. For a total synthesis of human knowledge, use the interrogative.

The conceptual artist/philosopher James Lee Byars contacted me and suggested a collaboration of sorts which resulted in our taking daily walks in Central Park as Byars and I walked and talked, convers-

ing only in interrogative sentences. Does it sound like
fun? Want to try it?

James Lee soon began to develop his ideas, which led to
"The World Question Center":

> To arrive at an axiology of the world's knowledge,
> seek out the most complex and sophisticated minds,
> put them in a room together, and have them ask each
> other the questions they are asking themselves.

On November 26, 1968, he launched "The World Question Center" in a one-hour television program produced in
Brussels at the studios of the Belgian national television
network and broadcast live to a national audience. During
the hour, he called numerous celebrated intellectuals such
as composer John Cage, science fiction writer Arthur C.
Clarke, futurist Herman Kahn, artist Joseph Beuys, novelist Jerzy Kosinski, poet Michael McClure, and asked, in
various ways, the following:

> I'm trying to find hypotheses that people are working
> with that are reduced into some type of very simple

single question with no explanation, hopefully, that's important to them in their own evolution of knowledge. Might you offer one that's personal?

For the 50th anniversary of "The World Question Center," and for the finale to the twenty years of *Edge* Questions, I turned it over to the *Edgies*:

"Ask 'The Last Question,' your last question, the question for which you will be remembered."

—John Brockman
Editor, *Edge*

Introduction Endnotes

1 John Brockman, *By the Late John Brockman* (New York: Macmillan, 1969).
2 Ludwig Wittgenstein, *Zettel*, eds. G. E. M. Anscombe and G. H. von Wright, trans. G. E. M. Anscombe (Berkeley: University of California Press, 1967), p. 58e, para. 315.
3 Alfred North Whitehead, *Process and Reality* (New York: Harper & Row, 1960), p.17.
4 Paul Valery, *The Outlook for Intelligence* (New York: Harper & Row, 1962).
5 Alain Robbe-Grillet, *For a New Novel*, trans. Richard Howard (New York: Grove Press, 1965).

The Last
Unknowns

Can we program

a computer

to find a

10,000-bit string

that encodes more

actionable wisdom

than any human

has ever expressed?

SCOTT AARONSON

David J. Bruton Centennial Professor of Computer
Science, University of Texas at Austin; author,
Quantum Computing Since Democritus

Are complex biological neural systems fundamentally unpredictable?

ANTHONY AGUIRRE

Professor of physics, University
of California, Santa Cruz

Are the simplest
bits of information
in the brain
stored at the level
of the neuron?

DORSA AMIR

Postdoctoral research fellow,
Boston College

Are people
who cheat
vital to
driving progress
in human
societies?

ALUN ANDERSON

Senior consultant (and former editor-in-chief
and publishing director), *New Scientist*;
author, *After the Ice*

How can

we put

rational prices

on human lives

without becoming

inhuman?

CHRIS ANDERSON

Chief executive officer,
3D Robotics;
founder, DIY Drones

How will
we build the tools
to maintain
the software in
long-lived online devices
that can kill us?

ROSS ANDERSON

Professor of security engineering,
University of Cambridge

How do
we best build
a civilization
that is galvanized
by long-term thinking?

SAMUEL ARBESMAN

Complexity scientist;
scientist in residence, Lux Capital;
author, *Overcomplicated*

How would changes in the marginal tax rate affect our efforts and motivation?

DAN ARIELY

James B. Duke Professor of Psychology and
Behavioral Economics, Duke University;
founding member,
Center for Advanced Hindsight;
author, *Predictably Irrational*

Will it ever
be possible
for us to transcend
our limited experience
of time as linear?

NOGA ARIKHA

Historian of ideas;
author, *Passions and Tempers*

Does consciousness reside only in our brains?

W. BRIAN ARTHUR

External professor, Santa Fe Institute; visiting researcher, Intelligent Systems Lab, Palo Alto Research Center

How can science best leverage unreason to overcome the heroic passion for war?

SCOTT ATRAN

Anthropologist; research director,
Centre National de la Recherche Scientifique,
Institut Jean Nicod;
cofounder, Centre for the Resolution
of Intractable Conflict, University of
Oxford; author, *Talking to the Enemy*

What is
the optimal
algorithm
for discovering
truth?

JOSCHA BACH

Cognitive scientist,
MIT Media Lab,
and Program for
Evolutionary Dynamics,
Harvard University

Will the
appearance of
new species of talented
computational intelligence
result in improving
the moral behavior
of persons and societies?

MAHZARIN BANAJI

Department chair and Richard Clarke Cabot Professor
of Social Ethics, Department
of Psychology, Harvard University;
coauthor (with Anthony Greenwald), *Blindspot*

Can we
re-design
our education system
based on
the principle
of neurodiversity?

SIMON BARON-COHEN

Professor of developmental psychopathology;
fellow, Trinity College;
director, Autism Research Centre,
University of Cambridge;
author, *Zero Degrees of Empathy*

How does
a single
human brain
architecture create
many kinds of
human minds?

LISA FELDMAN BARRETT

University Distinguished Professor of
Psychology, Northeastern University;
research neuroscientist,
Massachusetts General Hospital;
lecturer in psychiatry, Harvard Medical
School; author, *How Emotions Are Made*

What would
a diagram
that gave a
complete understanding
of imagination
need to be?

ANDREW BARRON

Neuroethologist,
Department of Biological Sciences,
Macquarie University

What libraries
will we have
to build when
cloning becomes
infinitely expandable?

THOMAS A. BASS

Professor of English and journalism,
State University of New York at Albany;
author, *The Spy Who Loved Us*

Will the process
of discovery
be completed
in any of
the natural sciences?

MARY CATHERINE BATESON

Professor emerita of anthropology, George Mason University; visiting scholar, Sloan Center on Aging & Work, Boston College; author, *Composing a Further Life*

What
is the
hard limit
on human
longevity?

GREGORY BENFORD

Professor emeritus
of physics and astronomy,
University of California, Irvine;
author, *The Berlin Project*

Will we
ever live
together
in a hive?

LAURA BETZIG

Anthropologist, historian

What kinds
of minds
could solve
the mind-body
problem?

SUSAN BLACKMORE

Psychologist; visiting professor, University
of Plymouth; author, *Consciousness*

Will
AI make
the
Luddites
(mostly) right?

ALAN S. BLINDER

Gordon S. Rentschler Memorial Professor
of Economics and Public Affairs,
Princeton University;
author, *Advice and Dissent*

Why are we
so often kind
to strangers
when nobody
is watching
and we have
nothing to gain?

PAUL BLOOM

Brooks and Suzanne Ragen Professor
of Psychology and Cognitive Science,
Yale University;
author, *Against Empathy*

How much
biodiversity
do we need?

GIULIO BOCCALETTI

Chief strategy officer,
The Nature Conservancy

Is there a way
for humans
to directly experience
what it's like
to be another entity?

IAN BOGOST

Ivan Allen College Distinguished Chair
in Media Studies and Professor of Interactive
Computing, Georgia Institute of Technology;
founding partner, Persuasive Games;
contributing editor, *The Atlantic*;
author, *Play Anything*

Will a machine
ever be able
to feel what
an organism feels?

JOSHUA BONGARD

Cyril G. Veinott Green and Gold Professor,
Department of Computer Science,
University of Vermont;
author, *How the Body Shapes the Way We Think*

Which questions should we not ask and not try to answer?

NICK BOSTROM

Professor of applied ethics, University of Oxford;
director, Future of Humanity Institute;
author, *Superintelligence*

Can wild animals
that are large
and dangerous
be made averse
to threatening
humans?

STEWART BRAND

Founder, *Whole Earth Catalog*;
cofounder, The WELL,
The Long Now Foundation,
and Revive & Restore;
author, *Whole Earth Discipline*

Can consciousness exist in an entity without a self-contained physical body?

RODNEY A. BROOKS

Panasonic Professor (emeritus)
of Robotics, Massachusetts Institute of Technology;
former director, MIT Artificial Intelligence
Laboratory and MIT Computer
Science & Artificial Intelligence Laboratory;
author, *Flesh and Machines*

Will scientific advances about the causes of sexual conflict help to end the "battle of the sexes"?

DAVID M. BUSS

Professor of psychology,
University of Texas at Austin;
author, *The Dangerous Passion*

How do I describe
the achievements,
meanings,
and power
of Beethoven's piano
sonata *Appassionata*?

PHILIP CAMPBELL

Editor-in-chief,
Springer Nature's portfolio of journals,
books, and magazines

When will we accept
that the most accurate clocks
will have to advance
regularly sometimes,
irregularly most of the time,
and at times run
counterclockwise?

JIMENA CANALES

Writer; faculty, Graduate College,
University of Illinois at Urbana-Champaign;
author, *The Physicist and the Philosopher*

How complex
must be the
initial design of the
simplest machine that
can learn from experience
to achieve, at a minimum,
the intelligence and abilities
of a typical human being?

CHRISTOPHER CHABRIS

Senior investigator, Geisinger Health System;
visiting fellow, Institute for Advanced Study;
coauthor (with Daniel Simons), *The Invisible Gorilla*

How can
we design
a machine that
can correctly answer
every question,
including this one?

DAVID CHALMERS

Professor of philosophy and neural science;
codirector, Center for Mind, Brain,
and Consciousness, New York University

What new methodology will be required to explain the neural basis of consciousness?

LEO M. CHALUPA

Neurobiologist;
professor of pharmacology and physiology,
School of Medicine and Health Sciences,
George Washington University

Is there a fundamental difference between the biological world and the physical world?

ASHVIN CHHABRA

Investor;
physicist;
author, *The Aspirational Investor*

41

Can we design
a modern society
without money
which is at least
as effective economically
and politically as
our current system?

JAEWEON CHO

Professor of environmental engineering,
Ulsan National Institute
of Science and Technology;
director, Science Walden Center

Will some things
about life,
consciousness,
and society
necessarily remain
unseen?

NICHOLAS A. CHRISTAKIS

Physician; social scientist;
Sterling Professor of Social and
Natural Science, Yale University;
coauthor (with James Fowler), *Connected*

Is the
unipolar future
of a "singleton"
the inevitable destiny
of intelligent life?

BRIAN CHRISTIAN

Author, *The Most Human Human*;
coauthor (with Tom Griffiths),
Algorithms to Live By

Will we

pass our audition

as planetary managers?

DAVID CHRISTIAN

Director,
Big History Institute,
and Distinguished Professor
in History, Macquarie University;
author, *Origin Story*

What will
we do as an encore
once we manage to
develop technological
solutions to infection,
aging, poverty,
asteroids, and heat
death of the universe?

GEORGE CHURCH

Robert Winthrop Professor of Genetics,
Harvard Medical School;
professor of health sciences and technology,
Harvard and Massachusetts Institute of Technology;
director, Personal Genome Project;
coauthor (with Ed Regis), *Regenesis*

Will we soon
cease to care
whether we are
experiencing normal,
augmented, or virtual reality?

ANDY CLARK

Professor of logic and metaphysics,
University of Edinburgh;
author, *Surfing Uncertainty*

What would comprise the most precise and complete sonic representation of the history of life?

JULIA CLARKE

John A. Wilson Professor in
Vertebrate Paleontology and
Howard Hughes Medical Institute Professor,
Jackson School of Geosciences,
University of Texas at Austin

How far
are we from
wishing to return
to the technologies
of the year
1900?

TYLER COWEN

Holbert L. Harris Chair of Economics;
distinguished senior fellow,
F. A. Hayek Program for
Advanced Study in Philosophy,
Politics, and Economics;
faculty director, Mercatus Center,
George Mason University

If science does in fact
confirm that we lack free will,
what are the implications
for our notions of blame,
punishment, reward, and
moral responsibility?

JERRY A. COYNE

Professor emeritus,
Department of Ecology and Evolution,
University of Chicago;
author, *Faith versus Fact*

Why do we
experience feelings
of meaning
in a universe
without purpose?

JAMES CROAK

Artist

Is technology
changing the nature
of moral emotions?

MOLLY CROCKETT

Assistant professor
of psychology, Yale University;
distinguished research fellow,
Oxford Centre for Neuroethics

Can natural selection's legacy of sex differences in values be reconciled with the universal values of the Enlightenment?

HELENA CRONIN

Codirector, Centre for Philosophy of Natural and Social Science, London School of Economics; author, *The Ant and the Peacock*

Why
be good?

OLIVER SCOTT CURRY

Senior researcher and director,
The Oxford Morals Project,
Institute of Cognitive and
Evolutionary Anthropology,
University of Oxford

Could the
thermodynamic prophecy
of an increasingly
entropic universe
be fulfilled by the
cosmic flourishing
of intelligent life?

DAVID DALRYMPLE

Research affiliate,
MIT Media Lab

What future
progressive norms
would most forward-thinking
people today dismiss as
too transgressive?

KATE DARLING

Research specialist,
MIT Media Lab;
affiliate, Berkman Klein Center
for Internet and Society
at Harvard University

If we want to make
a real and effective
science-based policy,
should we change
politics or science?

LUCA DE BIASE

Journalist;
editor, *Nòva—Il Sole 24 Ore*

Is our brain
fundamentally limited
in its ability
to understand
the external world?

STANISLAS DEHAENE

Neuroscientist;
professor of experimental cognitive psychology,
Collège de France;
author, *Consciousness and the Brain*

How can an aggregation of trillions of selfish, myopic cells discover the unwitting teamwork that turns that dynamic clump into a person who can love, notice, wonder, and keep a promise?

DANIEL C. DENNETT

Philosopher;
Austin B. Fletcher Professor
of Philosophy and Codirector,
Center for Cognitive Studies,
Tufts University;
author, *From Bacteria to Bach and Back*

Are accurate
mathematical
theories of
individual
human behavior
possible?

EMANUEL DERMAN

Professor of professional practice and director,
Master Financial Engineering Program,
Columbia University;
author, *Models. Behaving. Badly.*

Are the ways
qualia relate to computation,
creativity to free will,
risk to probability,
morality to epistemology,
all the same
question?

DAVID DEUTSCH

Visiting professor of physics, Centre for Quantum
Computation, University of Oxford;
author, *The Beginning of Infinity*;
recipient, *Edge* of Computation Science Prize

Can we develop
a procedure that, in principle,
would tell us whether or not
our universe is a simulation
(analogous to the way the now
proven Poincaré conjecture can
tell us the universe's shape)?

KEITH DEVLIN

Mathematician; cofounder and executive director,
Human-Sciences and Technologies
Advanced Research Institute,
Stanford University;
author, *Finding Fibonacci*

Why is there
such widespread
public opposition to
science and scientific
reasoning in
the United States,
the world leader
in every major
branch of science?

JARED DIAMOND

Professor of geography,
University of California, Los Angeles;
author, *The World Until Yesterday*

Will a computer ever really understand and experience human kindness?

CHRIS DIBONA

Director, Open Source and Making and Science Initiative, Google

Does this question exist in a parallel universe?

ROLF DOBELLI

Founder, Zurich Minds;
journalist;
author, *The Art of Thinking Clearly*

How will
we know
if we achieve
universal happiness?

P. MURALI
DORAISWAMY

Professor of psychiatry,
Division of Translational Neuroscience,
Duke University Health System

Is it ultimately possible
for life to bend the
shape of the universe
to fit life's purposes,
as we are now bending
the shape of our environment
here on earth?

FREEMAN DYSON

Physicist; professor emeritus of physics,
Institute for Advanced Study at Princeton;
author, *Maker of Patterns*

Why are
there
no trees
in the ocean?

GEORGE DYSON

Science historian;
author, *Turing's Cathedral*

Can we create
new senses for humans—
not just touch, taste,
vision, hearing, smell,
but totally novel qualia
for which we don't
yet have words?

DAVID M. EAGLEMAN

Neuroscientist and adjunct professor,
Stanford University; coauthor (with
Anthony Brandt), *The Runaway Species*

Will we ever be replaced
by another earthly species
capable of evolving to a
similar degree of social and
technical sophistication that
effectively fills the biocultural
niche we vacated?

DAVID EDELMAN

Neuroscientist;
visiting scholar, Dartmouth College

Is the cumulation of shared knowledge forever constrained by the limits of human language?

NICK ENFIELD

Professor and chair,
Department of Linguistics,
University of Sydney;
author, *How We Talk*

Have we
left the
Age of Reason,
never to
return?

BRIAN ENO

Artist;
composer;
recording producer for U2,
Coldplay, Talking Heads, Paul Simon;
recording artist

So,

before

the Singularity . . . ?

JUAN ENRIQUEZ

Managing director,
Excel Venture Management;
coauthor (with Steve Gullans), *Evolving Ourselves*

Will civilization collapse before I die?

DYLAN EVANS

Founder and chief executive officer,
Projection Point;
author, *The Utopia Experiment*

Will
humans
ever embrace
their own
diversity?

DANIEL L. EVERETT

Linguistic researcher;
dean of Arts and Sciences,
Bentley University;
author, *How Language Began*

Is there

a place

for our past

in our future?

CHRISTINE FINN

Archaeologist;
journalist;
author, *Past Poetic*

How many
incommensurable ideas
can we hold
in our mind
simultaneously?

STUART FIRESTEIN

Professor and chair, Department
of Biological Sciences, Columbia University;
fellow, American Association
for the Advancement of Science

What will courtship,
mate selection,
length of marriages,
and family composition
and networks be like
when we are all living
over 150 years?

HELEN FISHER

Biological anthropologist,
Rutgers University;
author, *Why Him? Why Her?*

Can we design
a common test
to assess machine,
animal, and human
intelligence?

STEVE FULLER

Philosopher; Auguste Comte Chair in Social
Epistemology, University of Warwick;
author, *Post-Truth*

Will the "third culture"
be followed by a fourth culture,
a fifth culture, and,
ominously, a Final Culture?

HOWARD GARDNER

John H. and Elisabeth A. Hobbs
Professor of Cognition and Education,
Harvard Graduate School of Education;
author, *Truth, Beauty, and Goodness Reframed*

Is there a single,
evolved biological mechanism
that can be tweaked
to improve overall health,
cognitive abilities,
and slow aging?

DAVID C. GEARY

Curators' Professor, Department
of Psychological Sciences,
University of Missouri-Columbia

Why is the phenomenon too familiar to investigate the hardest thing to completely understand?

JAMES GEARY

Deputy curator, Nieman Foundation
for Journalism at Harvard University;
author, *Wit's End*

Is
intersubjectivity
possible in a
quantum mechanical
universe?

AMANDA GEFTER

Physics writer;
author, *Trespassing on Einstein's Lawn*

Is there a
Turing test
for living rather
than thinking that
can distinguish animate
from automata?

NEIL GERSHENFELD

Physicist;
director, Center for Bits and Atoms, MIT;
coauthor (with Alan Gershenfeld
and Joel Cutcher-Gershenfeld), *Designing Reality*

What behaviors
are we attributing
only to brain mechanisms
that may be better explained
by considering biomechanics?

ASIF A. GHAZANFAR

Professor of neuroscience,
psychology, and ecology and
evolutionary biology, Princeton University

Is there
a single
theory of all physics,
and *what is it*?

STEVE GIDDINGS

Theoretical physicist;
professor, Department of Physics,
University of California, Santa Barbara

Can
human intuition
ever be
reduced to
an algorithm?

GERD GIGERENZER

Psychologist;
director, Harding Center for
Risk Literacy, Max Planck Institute
for Human Development;
author, *Risk Savvy*

How much time
will pass between
the last minute
before artificial
superintelligence
and the first minute
after it?

BRUNO GIUSSANI

European director and global curator, TED

Can we
acquire
complete access
to our unconscious
minds?

JOEL GOLD

Psychiatrist; clinical associate professor of
psychiatry, New York University School of Medicine;
coauthor (with Ian Gold), *Suspicious Minds*

Is it

possible to

control a system

capable of evolving?

NIGEL GOLDENFELD

Physicist;
professor of physics,
Center for Advanced Study, director,
NASA Astrobiology Institute for Universal Biology,
University of Illinois at Urbana-Champaign

If we're not the agents
of ourselves (and it's hard
to see how we can be),
how can we make sense
of moral accountability
(and how can we live
coherently without it)?

REBECCA
NEWBERGER GOLDSTEIN

Philosopher; novelist;
recipient, 2014 National Humanities Medal;
author, *Plato at the Googleplex*

Is there a
subtle form
of consciousness
that operates
independent
of brain
function?

DANIEL GOLEMAN

Psychologist;
coauthor (with Richard Davidson), *Altered Traits*

How can the few pounds of grey goo between our ears let us make utterly surprising, completely unprecedented, and remarkably true discoveries about the world around us, in every domain and at every scale, from quarks to quasars?

ALISON GOPNIK

Psychologist, University of California, Berkeley;
author, *The Gardener and the Carpenter*

Will a comprehensive
mathematics of
human behavior
ever be created?

JOHN GOTTMAN

Psychologist;
cofounder, The Gottman Institute;
author, *The Seven Principles for Making Marriage Work*

Are
stories
bad
for us?

JONATHAN GOTTSCHALL

Distinguished fellow,
Department of English,
Washington & Jefferson College;
author, *The Storytelling Animal*

What knowledge
and know-how
are our descendants
at risk of forgetting
as our species passes
through future
evolutionary
bottlenecks?

WILLIAM GRASSIE

Interdisciplinary scholar;
author, *The New Sciences of Religion*

What will happen
to human love
when we can design
the perfect robot lover?

KURT GRAY

Associate professor of psychology,
University of North Carolina, Chapel Hill;
coauthor (with Daniel Wegner), *The Mind Club*

What ethical responsibilities will humans owe to AGI systems?

A. C. GRAYLING

Master of New College of the Humanities;
supernumerary fellow, St. Anne's College,
University of Oxford;
author, *War*

What new cognitive abilities
will we need to live
in a world of
intelligent machines?

TOM GRIFFITHS

Professor of psychology and cognitive science,
director, Computational Cognitive
Science Lab, Princeton University;
coauthor (with Brian Christian), *Algorithms to Live By*

What is the upper limit
for how malleable
the human mind
and our emotions
can actually be?

JUNE GRUBER

Assistant professor of psychology,
University of Colorado, Boulder

Why is it
so hard
to find
the truth?

JONATHAN HAIDT

Social psychologist;
Thomas Cooley Professor of
Ethical Leadership, New York University
Stern School of Business;
coauthor (with Greg Lukianoff),
The Coddling of the American Mind

What will be
the use of
99 percent of humanity
for the 1 percent?

DAVID HAIG

George Putnam Professor of Biology,
Harvard University;
author, *Genomic Imprinting and Kinship*

Is scientific
knowledge the
most valuable
possession
of humanity?

HANS HALVORSON

Professor of philosophy,
Princeton University

Why do we get to ask questions at all?

TIMO HANNAY

Founding managing director, SchoolDash; co-organizer, Science Foo Camp

How could one
last question
possibly be enough?

JUDITH RICH HARRIS

Independent investigator and theoretician;
author, *The Nurture Assumption*

Is the
actual all
that is
possible?

SAM HARRIS

Neuroscientist;
philosopher;
author, *Waking Up*

Which facets of life
will we never understand
once biological and
cultural diversity
has vanished?

DANIEL HAUN

Professor of early child development and
culture and director, Leipzig Research Center for
Early Child Development, Leipzig University

Will
reading and writing
survive given the
seduction of
video and audio?

MARTI HEARST

Computer scientist;
professor, School of Information,
University of California, Berkeley;
author, *Search User Interfaces*

What
does it mean
to be
human?

DIRK HELBING

Professor of computational
social science, ETH Zurich;
affiliate professor,
Delft University of Technology

When will
we replace governments
with algorithms?

CÉSAR HIDALGO

Associate professor,
Massachusetts Institute of Technology;
director, Collective Learning, MIT Media Lab;
author, *Why Information Grows*

Will a baby
grown from an embryo
constructed from
human stem cells
eventually become
a person?

ROGER HIGHFIELD

Director, external affairs,
Science Museum Group;
coauthor (with Martin Nowak),
SuperCooperators

What is the principle
that causes complex
adaptive systems
(life, organisms, minds, societies)
to spontaneously emerge
from the interaction of
simpler elements
(chemicals, cells, neurons,
individual humans)?

W. DANIEL HILLIS

Physicist; computer scientist;
cofounder, Applied Invention;
author, *The Pattern on the Stone*

Will humanity
eventually exhaust
the unknown?

MICHAEL HOCHBERG

Evolutionist;
research director,
Centre National de la Recherche Scientifique;
external professor, Santa Fe Institute;
research visitor, Institute for
Advanced Study Toulouse

Why is it
that the maximum information
we can pack into a region
of space does not depend
on the volume of the region,
but only on the area
that bounds it?

DONALD D. HOFFMAN

Professor of cognitive science, University of
California, Irvine; author, *Visual Intelligence*

What would
the mind of
a child raised
in total isolation
of other animals
be like?

BRUCE HOOD

Professor of developmental psychology
in society, University of Bristol;
founder, Speakezee;
author, *The Self-Illusion*

Does every
mathematical symmetry
have a
manifestation
in the
physical world?

DANIEL HOOK

Chief executive officer,
Digital Science

What will it take

to end war

once and for all?

JOHN HORGAN

Director,
Center for Science Writings,
Stevens Institute of Technology

How can we
separate the assessment
of scientific evidence
from value judgments?

SABINE HOSSENFELDER

Research fellow,
Frankfurt Institute
for Advanced Studies;
author, *Lost in Math*

Why
is the world
so beautiful?

NICHOLAS HUMPHREY

Professor emeritus of psychology,
London School of Economics;
visiting professor of philosophy,
New College of the Humanities;
senior member, Darwin College,
University of Cambridge;
author, *Soul Dust*

How does
a thought
become a feeling?

MARCO IACOBONI

Neuroscientist;
professor of psychiatry
and biobehavioral sciences,
David Geffen School of Medicine,
University of California,
Los Angeles;
author, *Mirroring People*

What is the
biological price
of being a species
with a sense of humor?

ISABEL
BEHNCKE IZQUIERDO

Primatologist; associate professor,
Research Center for Social Complexity,
Universidad del Desarrollo; visiting researcher,
Social and Evolutionary Neuroscience
Research Group, University of Oxford

When will
race disappear?

NINA JABLONSKI

Biological anthropologist and paleobiologist;
Evan Pugh University Professor of
Anthropology at Pennsylvania State University

Will humanity end up with one culture?

MATTHEW O. JACKSON

William D. Eberle Professor of Economics,
Stanford University;
external faculty member,
Santa Fe Institute; senior fellow, CIFAR

What systems
could be put
in place to prevent
widespread denial
of science-based
knowledge?

JENNIFER JACQUET

Assistant professor of
environmental studies,
New York University;
author, *Is Shame Necessary?*

Will the
"hard problem"
of consciousness dissolve
(rather than be solved)
as we learn more about
the natural world?

DALE JAMIESON

Professor of environmental studies and
philosophy, New York University

i = we ?

KOO JEONG-A

Artist

What might
the last fully
biological human's
statement be at
their last supper?

LORRAINE JUSTICE

Dean emerita and professor
of industrial design,
Rochester Institute of Technology

Are the laws
of physics unique
and inevitable?

GORDON KANE

Theoretical particle physicist and cosmologist;
Victor Weisskopf Distinguished University
Professor, University of Michigan;
author, *Supersymmetry and Beyond*

What

is

consciousness?

STUART A. KAUFFMAN

Professor of biological sciences, physics,
and astronomy, University of Calgary;
author, *Reinventing the Sacred*

Is there
any observational
evidence that
could shake
your faith,
or lack thereof?

BRIAN G. KEATING

Astrophysicist;
professor of physics,
University of California, San Diego;
author, *Losing the Nobel Prize*

Why don't naked mole rats age or get cancer?

PAUL KEDROSKY

Editor, Infectious Greed;
general partner, SK Ventures

How can
the process of science
be improved?

KEVIN KELLY

Senior maverick, *Wired*;
author, *The Inevitable*

How can we build machines that make us smarter?

GARY KLEIN

Senior scientist, MacroCognition; author, *Seeing What Others Don't*

Can we create
technologies that
help equitably
reduce the amount
of conflict
in the world?

JON KLEINBERG

Tisch University Professor of Computer Science
and Information Science, Cornell University

How can
we achieve
closed-loop
neural control
of human hedonics?

BRIAN KNUTSON

Professor of psychology and
neuroscience, Stanford University

What is the bumpiest
and highest-dimensional
cost surface that our
best computers will be able
to search and still find the
deepest cost well?

BART KOSKO

Information scientist and professor
of electrical engineering and law,
University of Southern California;
author, *Noise*

Can
brain implants
make us
better
human beings?

STEPHEN M. KOSSLYN

Founding dean, Minerva Schools at
the Keck Graduate Institute

Is our
continued coexistence
with the other
big mammals
essential to furthering
our understanding
of human cognition?

JOHN W. KRAKAUER

John C. Malone Professor of Neurology, Neuroscience,
and Physical Medicine and Rehabilitation, director,
Brain, Learning, Animation, and Movement Lab,
Johns Hopkins University School of Medicine

What will
happen to religion
on Earth
when the first
alien life form
is found?

KAI KRAUSE

Software pioneer; philosopher;
author, *A Realtime Literature Explorer*

Do we need
checks and balances
for virtual worlds?

ANDRIAN KREYE

Editor, *The Feuilleton* (Arts and Essays),
supplement of the German daily
newspaper *Sueddeutsche Zeitung*

Why do we
care so much
about how well
we're approximated
by algorithms?

COCO KRUMME

Applied mathematician,
University of California, Berkeley;
founder, Leeward Co.

Has consciousness done more good or bad for humanity?

JOSEPH LEDOUX

Henry and Lucy Moses Professor of Science,
professor of psychiatry and child and
adolescent psychiatry, New York University;
director, Emotional Brain Institute;
author, *Anxious*

Will human psychology
keep pace with
the exponential growth
of technological innovation
associated with
cultural evolution?

CRISTINE H. LEGARE

Associate professor of psychology,
The University of Texas at Austin;
director, Evolution, Variation, and
Ontogeny of Learning Laboratory

What proportion
of
"ethnic" and "religious"
tensions are rooted
in our genes?

MARTIN LERCHER

Professor of computational cell biology,
Heinrich Heine University;
coauthor (with Itai Yanai), *The Society of Genes*

Are humans capable of building a moral economy?

MARGARET LEVI

Sara Miller McCune Director,
Center for Advanced Study in the Behavioral Sciences,
and Professor of Political Science, Stanford University;
Jere L. Bacharach Professor Emerita of
International Studies, University of Washington

Is gravity
a fundamental
law of nature,
or does gravity—and
thereby spacetime—
emerge as a
consequence of the
underlying quantum
nature of reality?

JANNA LEVIN

Professor of physics and astronomy,
Barnard College, Columbia University;
author, *Black Hole Blues and Other
Songs from Outer Space*

Where were the laws
of physics written
before the universe
was born?

ANDREI LINDE

Theoretical physicist;
professor of physics, Stanford University;
father of eternal chaotic inflation;
inaugural recipient, Fundamental Physics Prize

What is the fundamental geometric structure underlying reality?

ANTONY GARRETT LISI

Theoretical physicist

Will it ever
be possible to
download the information
stored in the human brain?

MARIO LIVIO

Astrophysicist;
author, *Why?*

How did
our complex universe
arise out of
simple physical laws?

SETH LLOYD

Professor of quantum mechanical engineering,
Massachusetts Institute of Technology;
author, *Programming the Universe*

How will advances
in mental prosthetics
that connect us with other
human and machine minds
change the way we think
about expertise?

TANIA LOMBROZO

Professor of psychology,
director, Concepts and Cognition
Lab, Princeton University

How will evolution shape the biological world one hundred years from now, or one hundred thousand?

JONATHAN B. LOSOS

Monique and Philip Lehner Professor
for the Study of Latin America and
Professor of Organismic and
Evolutionary Biology, Harvard University;
curator in herpetology,
Museum of Comparative Zoology;
author, *Improbable Destinies*

Can we train machines
to design and construct
a humane and vibrant built
environment for us?

GREG LYNN

Owner, Greg Lynn FORM;
Ordentlicher University Professor of Architecture,
University of Applied Arts Vienna;
studio professor, University of California,
Los Angeles School of the Arts and Architecture

How will people focus
more on forming the
right question,
before rushing headlong
toward the answer?

ZIYAD MARAR

President of global publishing, SAGE;
author, *Judged*

Why are humans still so much more flexible in their thinking and everyday reasoning than machines?

GARY MARCUS

Professor of psychology,
New York University;
director, NYU Center for
Language and Music;
author, *Guitar Zero*

How will the world
be changed when
battery storage technology
improves at the same
exponential rate seen
in computer chips
in recent decades?

JOHN MARKOFF

Pulitzer Prize–winning reporter,
The New York Times;
author, *Machines of Loving Grace*

Is the number
of interesting questions
finite or not?

CHIARA MARLETTO

Junior research fellow, Wolfson College, and
postdoctoral research associate, Materials
Department, University of Oxford

When in the evolution
of animal life did the capacity
to experience love for another
being first emerge?

ABIGAIL MARSH

Associate professor of psychology,
Georgetown University

How much
of what we
call "reality" is
ultimately grounded
and instantiated
in convincing
communication
and storytelling?

BARNABY MARSH

Evolutionary dynamics scholar,
Program for Evolutionary Dynamics,
Harvard University;
visitor, Institute for Advanced Study

What is the
master principle
governing the
growth and evolution
of complex systems?

JOHN C. MATHER

Recipient, 2006 Nobel Prize in Physics;
senior astrophysicist, Observational Cosmology
Laboratory, Goddard Space Flight Center, NASA

Why
are people
so seldom persuaded
by
clear evidence
and
rational argument?

TIM MAUDLIN

Professor of philosophy,
New York University

Is love

really

all you need?

ANNALENA McAFEE

Journalist;
founding editor, *The Guardian Review*,
literary supplement *of The Guardian*;
author, *Hame*

Are humans
ever really capable
of regarding others
as ends in themselves?

MICHAEL McCULLOUGH

Professor of psychology and director,
Evolution and Human Behavior Laboratory,
University of Miami;
author, *Beyond Revenge*

If the sum
of all significant
knowledge
is finite, what
proportion
of it can humans,
aided by
intelligent machines,
eventually attain?

IAN McEWAN

Novelist;
author, *Atonement, Solar,
On Chesil Beach, Nutshell*

Will we be
one of the last
generations
in human history
that dies?

RYAN McKAY

Professor of psychology, Royal Holloway,
University of London

Can major historical events,
from the advent of moral religions
to the Industrial Revolution,
be explained by changes in
life history strategies?

HUGO MERCIER

Cognitive scientist,
Centre National de la Recherche Scientifique;
coauthor (with Dan Sperber), *The Enigma of Reason*

What is the most intelligent
and efficient way to minimize
the overall amount of conscious
suffering in the universe?

THOMAS METZINGER

Professor of theoretical philosophy,
Johannes Gutenberg-Universität Mainz;
adjunct fellow, Frankfurt Institute
for Advanced Studies;
author, *The Ego Tunnel*

If we
discover another
intelligent civilization,
what should we
ask them?

YURI MILNER

Physicist; entrepreneur and
venture capitalist;
science philanthropist

Are feelings
computable?

READ MONTAGUE

Neuroscientist; professor and director,
Human Neuroimaging Laboratory
and Computational Psychiatry Unit,
Virginia Tech Carilion Research Institute;
author, *Why Choose This Book?: How We Make Decisions*

Is the brain
a computer
or an antenna?

DAVE MORIN

Internet entrepreneur,
angel investor

Is there an evolutionary advantage to building societies that favor entertaining over understanding?

LISA MOSCONI

Associate professor of neuroscience and associate director of the Alzheimer's Prevention Clinic at Weill Cornell Medical College/NewYork-Presbyterian Hospital; author, *Brain Food*

Does religious engagement promote or impede morality, altruism, and human flourishing?

DAVID G. MYERS

Professor of psychology,
Hope College

Are there
limits to what
we can know
about the
universe?

PRIYAMVADA NATARAJAN

Theoretical astrophysicist; professor,
Departments of Astronomy and Physics,
Yale University;
author, *Mapping the Heavens*

Why do humans
who possess or
acquire unaccountable power
over others invariably
abuse it?

JOHN NAUGHTON

Senior research fellow, Centre for Research
in the Arts, Social Sciences and Humanities,
University of Cambridge; director,
Wolfson College Press Fellowship Programme;
columnist, *The Observer*;
author, *From Gutenberg to Zuckerberg*

In what situations
does the capacity
for low mood
give a selective
advantage?

RANDOLPH NESSE

Professor of life sciences and founding director,
Center for Evolution and Medicine,
Arizona State University;
coauthor (with George C. Williams),
Why We Get Sick

What does
the conscious mind
do that is impossible
for the unconscious mind?

RICHARD NISBETT

Theodore M. Newcomb Distinguished
University Professor of Psychology,
University of Michigan;
author, *Mindware*

What is the flow
of information
through human beings?

TOR NØRRETRANDERS

Writer, speaker, thinker

Why do humans
behave as though
what can be known
is finite?

MICHAEL I. NORTON

Harold M. Brierley Professor of
Business Administration and
Director of Research, Harvard Business School;
coauthor (with Elizabeth Dunn), *Happy Money*

What is
the purpose of it?

MARTIN NOWAK

Professor of biology and mathematics and director,
Program for Evolutionary Dynamics,
Harvard University;
coauthor (with Roger Highfield),
SuperCooperators

Does the future belong to nonhuman entities?

HANS ULRICH OBRIST

Curator, Serpentine Galleries;
editor, *A Brief History of Curating*;
coauthor (with Rem Koolhaas), *Project Japan*

When will
"human being"
cease to be a
meaningful category
to speak of?

JAMES J. O'DONNELL

Classics scholar and university librarian,
Arizona State University; author, *Pagans*

How did
our sense
of mathematical
beauty arise?

STEVE OMOHUNDRO

Scientist, Self-Aware Systems;
cofounder, Center for Complex Systems Research

What can humanity
do right now
that will make
the biggest difference
over the next billion years?

TOBY ORD

Philosopher, University of Oxford;
research fellow, Future of Humanity Institute

How can AI and
other digital technologies
help us create
global institutions
that we can trust?

TIM O'REILLY

Founder and chief executive officer,
O'Reilly Media; author, *WTF?*

Why do even the
most educated people
today feel that their grip
on what they can truly know
is weaker than ever before?

GLORIA ORIGGI

Philosopher and researcher,
Centre National de la Recherche Scientifique;
author, *Reputation*

Is a single world language and culture inevitable?

MARK PAGEL

Professor of evolutionary biology,
University of Reading;
fellow, Royal Society;
author, *Wired for Culture*

Why is religion still around in the twenty-first century?

ELAINE PAGELS

Harrington Spear Paine
Professor of Religion, Princeton University;
author, *Why Religion?*

Is the assertion
"Nothingness is impossible"
the most fundamental statement
we can make about our existence?

BRUCE PARKER

Visiting professor,
Stevens Institute of Technology;
author, *The Power of the Sea*

Can we engineer a human being?

JOSEF PENNINGER

Director, Institute of Molecular Biotechnology,
Austrian Academy of Sciences;
director, Life Science Institute,
University of British Columbia;
professor of genetics, University of Vienna

What is the
most important thing
that can be done
to restore the general
public's faith and trust in science?

IRENE PEPPERBERG

Research associate and lecturer,
Harvard University;
adjunct associate professor,
Brandeis University;
author, *Alex & Me*

Will humans
ever prove the
Riemann hypothesis
in mathematics?

CLIFFORD PICKOVER

Author, *The Math Book*

How
can we
empower the
better angels
of our
nature?

STEVEN PINKER

Johnstone Family Professor of Psychology,
Harvard University;
author, *Enlightenment Now*

Are we smart enough
to know when we've reached
the limits of our ability
to understand the universe?

DAVID PIZARRO

Associate professor of psychology,
Cornell University

How far will we go
in predicting
human behavior
from DNA?

ROBERT PLOMIN

Professor of behavioral genetics,
King's College London

Will blockchain
return us to the golden age
of ownership of
information licenses
that can be resold
like books and records?

JORDAN POLLACK

Professor of complex systems and chairman of
computer science, Brandeis University

Will artistic invention enlighten the age of AI?

ALEX POOTS

Artistic director and chief executive officer,
The Shed

What will
it take for
us to be fully
confident that we
have found life
elsewhere in the cosmos?

CAROLYN PORCO

Planetary scientist;
Cassini Imaging Team Leader and director,
CICLOPS, Space Science Institute

Does the infinite multiverse
of cosmologists, in which all that
is physically possible occurs,
contain realizations of
our unruly paradoxes of infinity
(Hilbert's Hotel, Thomson's lamp,
$1+2+3+4 \ldots = -\frac{1}{12}$; etc.)?

WILLIAM POUNDSTONE

Journalist;
author, *Head in the Cloud*;
two-time Pulitzer Prize nominee

What quirk of evolution caused us to develop the ability to do pure mathematics?

WILLIAM H. PRESS

Warren J. and Viola Mae Raymer Chair,
Department of Computer Science and School
of Biological Sciences, Integrative Biology
Section, University of Texas at Austin

Can a single underlying process
explain the emergence of
structure at the physical,
biological, cognitive,
and machine levels?

ROBERT R. PROVINE

Research professor and professor emeritus,
University of Maryland, Baltimore County;
author, *Curious Behavior*

Will questioning
be replaced
by answering
without questions?

MATTHEW PUTMAN

Applied physicist; chairman,
board of directors, Pioneer Works;
chief executive officer, Nanotronics

Must we suffer
and die?

DAVID C. QUELLER

Evolutionary biologist,
Washington University in St. Louis

Will weaving networks that blend humans and machines yield network effects?

SHEIZAF RAFAELI

Professor and director,
The Center for Internet Research,
University of Haifa

Why should we prize the original object over a perfect replica?

VILAYANUR RAMACHANDRAN

Neuroscientist; professor and director, Center for Brain and Cognition, University of California, San Diego; author, *The Tell-Tale Brain*

How far
can we extend
beyond our
human limitations
to more fully grasp
the nature of
the world?

LISA RANDALL

Physicist and Frank B. Baird Jr. Professor
of Science, Harvard University;
author, *Dark Matter and the Dinosaurs*

Why
is sleep
so necessary?

S. ABBAS RAZA

Founding editor,
3 Quarks Daily

Will it be possible to do surgical operations in the future without making incisions?

SYED TASNIM RAZA

Associate professor of surgery, Columbia University Medical Center; medical director, Cardiac Surgery Step-Down Unit, Columbia University Medical Center and NewYork-Presbyterian Hospital

Will
post-humans
be organic
or electronic?

MARTIN REES

Former president, Royal Society;
fellow, Trinity College;
emeritus professor of cosmology
and astrophysics, University of Cambridge;
author, *From Here to Infinity*

Why are reason, science, and evidence so impotent against superstition, religion, and dogma?

ED REGIS

Science writer;
author, *Monsters*

How smart
does another animal
have to be for us to
decide not to eat it?

DIANA REISS

Professor of psychology, Hunter College;
author, *The Dolphin in the Mirror*

What
does justice
feel like?

JENNIFER RICHESON

Philip R. Allen Professor of Psychology,
Yale University

Is the botscape
going to force us
to give up the use of the
first-person singular
nominative case
personal pronoun, *I*?

GIANLUIGI RICUPERATI

Writer, essayist, curator

When will we
develop a robust
theory of
ontological intelligence?

MATTHEW RITCHIE

Artist

Can a user-friendly computer
proof assistant satisfy the
mathematician's desire
for certainty without
killing the pleasure?

SIOBHAN ROBERTS

Journalist-in-residence,
Max Planck Institute for
the History of Science, Berlin;
author, *Genius at Play*

Can
you
prove
it?

ANDRÉS ROEMER

**Curator and cofounder, La Ciudad de las Ideas;
coauthor (with Clotaire Rapaille), *Move UP***

How can aims
of individual liberty
and economic efficiency
be reconciled with aims
of social justice and
environmental sustainability?

PHIL ROSENZWEIG

Professor of strategy and international business,
IMD Switzerland;
author, *Left Brain, Right Stuff*

Is there an ultimate reality?

CARLO ROVELLI

Theoretical physicist and professor, Centre de Physique Théorique, Aix-Marseille University; author, *Reality Is Not What It Seems*

Was agriculture
a wrong turn
for civilization?

DOUGLAS RUSHKOFF

Media analyst; documentary writer;
professor of media theory and digital economics,
Queens College, City University of New York;
author, *Team Human*

How will humanity
change in light
of the increasing use
of nonsexual methods
of reproduction?

KARL SABBAGH

Writer and television producer;
author, *Remembering Our Childhood*

Will there ever be
a mechanistic scientific question
that can be asked about the
lone individuality of mental life,
with its particular beginning,
middle, and end?

TODD C. SACKTOR

Distinguished Professor of Physiology,
Pharmacology, and Neurology,
State University of New York
Downstate Medical Center

Will we
ever be able
to predict earthquakes?

PAUL SAFFO

Technology forecaster;
consulting associate professor,
Stanford University

Why do
some people
act inside the law
and
others outside,
and others create
the law?

EDUARDO SALCEDO-ALBARAN

Philosopher;
director, Vortex Foundation

Can technology tame evolution?

BUDDHINI SAMARASINGHE

Molecular biologist;
founder, STEM Women

How do ideas
about biological evolution
change once one species
has control over the origin
and extinction
of all other species?

SCOTT SAMPSON

President and chief executive officer,
Science World British Columbia;
Dinosaur paleontologist and
science communicator;
author, *How to Raise a Wild Child*

What

cognitive capacities

make humans

so damn weird

relative to all the other

animals on the planet?

LAURIE R. SANTOS

Professor of psychology and director,
Comparative Cognition Laboratory and the
Canine Cognition Center, Yale University

Given the nature of life,

the purposeless indifference

of the universe,

and our complete lack of free will,

how is it that most people

avoid ever being clinically depressed?

ROBERT SAPOLSKY

Neuroscientist and John A. and Cynthia Fry
Gunn Professor and Professor of Neurology and of
Neurosurgery, Stanford University;
author, *Behave*

What is the
cosmic perspective
to the future of life?

DIMITAR D. SASSELOV

Phillips Professor of Astronomy and director,
Origins of Life Initiative, Harvard University;
author, *The Life of Super-Earths*

Why is it so difficult to
influence people's belief systems
for deeply held beliefs and so
easy to manipulate belief systems
when little is known
about the subject?

ROGER SCHANK

Chairman and chief executive officer, Socratic Arts;
John Evans Professor Emeritus of
Computer Science, Education,
and Psychology, Northwestern University

Is a human brain capable of understanding a human brain?

RENÉ SCHEU

Editor-in-chief,
Feuilleton supplement of *Neue Zürcher Zeitung*

What are
the beautiful curiosities
that artificial curiosities
can't comprehend?

MAXIMILIAN SCHICH

Associate professor for arts and technology,
The University of Texas at Dallas

Is immortality
desirable?

SIMONE SCHNALL

Director, Cambridge Body, Mind,
and Behavior Laboratory;
reader in experimental social psychology and
fellow, Jesus College, University of Cambridge

Can an increasingly
powerful species survive
(and overcome)
the actions of its
most extreme individuals?

BRUCE SCHNEIER

Fellow and lecturer in Public Policy,
Harvard Kennedy School;
author, *Data and Goliath*

Is the universe relatively
simple and comprehensible
by the human brain,
or is it so complex,
higher dimensional,
and multiversal that it
remains forever elusive
to humans?

PETER SCHWARTZ

Futurist; senior vice president for
global government relations and
strategic planning, Salesforce.com;
author, *Inevitable Surprises*

Can the human brain
ever fully understand
quantum mechanics?

GINO SEGRE

Professor of physics emeritus,
University of Pennsylvania;
author, *The Pope of Physics*

How can we reap the benefits
of the wide and open exchange
of data without undermining
the values that depend upon
the scarcity of information?

CHARLES SEIFE

Professor of journalism, New York University;
former journalist, *Science* magazine;
author, *Virtual Unreality*

How diverse
is life in the universe?

TERRENCE J. SEJNOWSKI

Francis Crick Chair,
Salk Institute for Biological Studies;
coauthor (with Patricia Churchland),
The Computational Brain

Would you like
to live a thousand years?

MICHAEL SHERMER

Publisher, *Skeptic* magazine;
monthly columnist, *Scientific American*;
presidential fellow, Chapman University;
author, *Heavens on Earth*

Will we ever find
an organization form
that brings out the best
in people?

OLIVIER SIBONY

Affiliate professor, HEC Paris

Is civilization's demand for water a dividing or unifying force?

LAURENCE C. SMITH

Professor of geography and
professor of earth, planetary,
and space sciences,
University of California, Los Angeles;
author, *The World in 2050*

Why is
human communication
embedded in the silence
of material objects?

MONICA L. SMITH

Navin and Pratima Doshi Chair in Indian Studies
and professor of anthropology,
University of California, Los Angeles

Why is the acceleration
of the expansion of the
universe roughly equal to
a typical acceleration of a star
in a circular orbit in a disk galaxy?

LEE SMOLIN

Physicist, Perimeter Institute;
author, *Time Reborn*

Does romantic love have a biological function?

DAN SPERBER

Distinguished visiting professor,
Central European University;
professor emeritus,
Centre National de la Recherche Scientifique;
coauthor (with Hugo Mercier), *The Enigma of Reason*

Could superintelligence be the purpose of the universe?

MARIA SPIROPULU

Shang-Yi Ch'en Professor of Physics,
California Institute of Technology;
founder, Alliance for Quantum Technologies,
Intelligent Quantum Networks & Technologies

What would
the ability to
synthesize creativity
do to cultural evolution?

NINA STEGEMAN

Associate editor, *Edge*

Will the universe observed today someday begin to contract, bounce, and be reborn?

PAUL J. STEINHARDT

Albert Einstein Professor in Science,
Departments of Physics and
Astrophysical Sciences, Princeton University;
author, *The Second Kind of Impossible*

Do the laws
of physics change
with the passage
of time?

BRUCE STERLING

Science fiction author, *Mirrorshades*

Why did we acquire our extraordinary human capacity for social learning?

STEPHEN J. STICH

Board of Governors Professor,
Department of Philosophy,
Rutgers University

How do I know the right level of abstraction at which to explain a phenomenon?

VICTORIA STODDEN

Associate professor of information sciences,
University of Illinois at Urbana-Champaign

Can we ever
wean humans
off their addiction
to religion?

CHRISTOPHER STRINGER

Paleoanthropologist;
author, *Lone Survivors*

Will our
AI future forms
need the
natural world?

SEIRIAN SUMNER

Reader in behavioral ecology,
University College London

Is there a design
to the laws of physics,
or are they the result
of chance and the laws
of large numbers?

LEONARD SUSSKIND

Felix Bloch Professor of
Theoretical Physics,
Stanford University;
author, The Theoretical Minimum series

Will the behavior
of a superintelligent AI
be mostly determined
by the results of its reasoning
about the other superintelligent AIs?

JAAN TALLINN

Cofounder, Centre for the Study
of Existential Risk, University of Cambridge;
cofounder, Future of Life Institute;
founding engineer, Skype and Kazaa

Why is
Homo sapiens
the sole nonextinct
species of hominin?

TIMOTHY TAYLOR

Professor of the prehistory of humanity,
University of Vienna;
author, *The Artificial Ape*

What will be
the literally last question
that will preoccupy
future superintelligent
cosmic life for as long as
the laws of physics permit?

MAX TEGMARK

Physicist and professor of physics,
Massachusetts Institute of Technology;
scientific director, Foundational Questions Institute;
cofounder, Future of Life Institute; author, *Life 3.0*

How will we cope when we are capable of keeping humans alive longer than our optimal life expectancy?

RICHARD H. THALER

Father of behavioral economics;
recipient, 2017 Nobel Memorial Prize
in Economic Sciences;
director, Center for Decision Research,
University of Chicago Booth School of Business;
author, *Misbehaving*

Can rational beings
such as Bayesian robots,
humans, and superintelligent
AIs ever reach agreement?

FRANK TIPLER

Professor of mathematical physics,
Tulane University;
author, *The Physics of Christianity*

Can behavioral science crack the ultimate challenge of getting people to durably adopt much healthier lifestyles?

ERIC TOPOL

Executive vice president, Scripps Research;
founder and director, Scripps Research
Translational Institute; professor of molecular
medicine; author, *The Patient Will See You Now*

What will time with
artifacts that simulate the
emotional experience of
being with another person
do to our human capacity
to handle the surely
rougher, more frictional,
and demanding human
intimacies on offer?

SHERRY TURKLE

Internet culture researcher;
Abby Rockefeller Mauzé Professor of the
Social Studies of Science and Technology,
Massachusetts Institute of Technology;
author, *Reclaiming Conversation*

How do the limits of the mind limit our understanding?

BARBARA TVERSKY

Professor of psychology and education,
Columbia Teachers College;
professor emerita of psychology,
Stanford University

How can coalitions of scholars who wish to update the content of explicit common knowledge in order to use that knowledge collaboratively detect and circumvent coalitions which are applying narrative control strategies to preserve arbitrage opportunities implicit in disparities between official narratives and reality?

MICHAEL VASSAR

Cofounder and chief science officer,
MetaMed Research

Will the creation of a superhuman class from a combination of genome editing and direct biological-machine interfaces lead to the collapse of civilization?

J. CRAIG VENTER

Founder, chairman, and chief executive office,
J. Craig Venter Institute;
cofounder and scientific advisor,
Synthetic Genomics;
cofounder, Human Longevity, Inc.;
author, *A Life Decoded*

Will we ever
understand how
human communication
is built from genes to cells
to circuits to behavior?

ATHENA VOULOUMANOS

Associate professor of psychology
and principal investigator, NYU Infant Cognition
and Communication Lab, New York University

Are there any
phenomena for which
it will never be
possible to develop
parsimonious theories?

D. A. WALLACH

Recording artist;
songwriter;
investor

Are moral beliefs
more like facts
or more like preferences?

ADAM WAYTZ

Psychologist; associate professor
of management and organizations,
Kellogg School of Management
at Northwestern University

Can humans
set a nonevolutionary course
that is game-theoretically stable?

BRET WEINSTEIN

Theoretical evolutionary biologist

Does something
unprecedented happen
when we finally learn our
own source code?

ERIC R. WEINSTEIN

Mathematician and economist;
managing director, Thiel Capital

How do we create
and maintain backup options
for humanity to
quickly rebuild an advanced
civilization after a catastrophic
human extinction event?

ALBERT WENGER

Managing partner, Union Square Ventures

How and when
will it end
or will it persist
indefinitely?

GEOFFREY WEST

Distinguished Professor and
past president, Santa Fe Institute;
author, *Scale*

How will the advent
of direct brain-to-brain communication
change the way we think?

THALIA WHEATLEY

Associate professor of psychological and
brain sciences, Dartmouth College

How much would surrendering our god(s) strengthen the odds of our survival?

TIM WHITE

Paleoanthropologist;
professor of integrative biology,
University of California, Berkeley

How can we sculpt
how individual brains
develop to avert
mental illness?

LINDA WILBRECHT

Associate professor of psychology,
Helen Wills Neuroscience Institute,
University of California, Berkeley

Why?

FRANK WILCZEK

Physicist;
Herman Feshbach Professor of Physics,
Massachusetts Institute of Technology;
recipient, 2004 Nobel Prize in Physics;
author, *A Beautiful Question*

Why are the errors
that our best
machine-learning algorithms
make so different
from the errors
we humans make?

JASON WILKES

Graduate student in psychology,
University of California, Santa Barbara;
author, *Burn Math Class*

What will be obvious
to us in a generation
that we have
an inkling of today?

EVAN WILLIAMS

Chief executive officer, Medium

Can general-purpose computers be constructed out of pure gravity?

ALEXANDER WISSNER-GROSS

Scientist, inventor, entrepreneur, investor

Can the pace
of human evolution
stop accelerating?

MILFORD H. WOLPOFF

Professor of anthropology,
University of Michigan;
adjunct associate research scientist,
Museum of Anthropology

In which
century or millennium
can all humanity
be expected to speak
the same primary language?

RICHARD WRANGHAM

Ruth B. Moore Professor of
Biological Anthropology,
Harvard University;
author, *Catching Fire*

How do our microbes
contribute to that
particular combination
of continuity and change
that makes us human?

ELIZABETH
WRIGLEY-FIELD

Assistant professor of sociology,
University of Minnesota-Twin Cities;
faculty member, Minnesota Population Center

Clarify the differences
between understanding,
knowledge, and wisdom
that could be communicated
to a literate twelve-year-old and
recommunicated to their parents.

RICHARD SAUL WURMAN

Founder, TED, EG, and
TEDMED conferences;
architect; cartographer;
author, *Information Architects*

How do contemporary developments in technology affect human cultural diversity?

VICTORIA WYATT

Associate professor of history in art,
University of Victoria

How can we rebel
against our genes
if we are biological creatures
without free will?

ITAI YANAI

Director, Institute for Computational Medicine,
and professor of biochemistry
and molecular pharmacology,
New York University School of Medicine;
coauthor (with Martin Lercher), *The Society of Genes*

Will the frontiers
of consciousness
be technological
or linguistic?

DUSTIN YELLIN

Artist;
founder, Pioneer Works

What is the fastest way
to reliably align a powerful AGI
around the safe performance
of some limited task that is
potent enough to save the world
from unaligned AGI?

ELIEZER S. YUDKOWSKY

Research fellow and cofounder,
Machine Intelligence Research Institute

What is
the world
without
the mind?

DAN ZAHAVI

Professor of philosophy and director,
Center for Subjectivity Research,
University of Copenhagen;
author, *Self and Other*

Will the individual quantum event forever remain random?

ANTON ZEILINGER

Physicist; professor emeritus of physics,
University of Vienna;
senior scientist, Institute of
Quantum Optics and Quantum Information;
president, Austrian Academy of Sciences;
author, *Dance of the Photons*

How does

the past

give rise

to the future?

CARL ZIMMER

New York Times journalist;
author, *She Has Her Mother's Laugh*

Index

About the Author

The founder and publisher of the online science salon
Edge.org, John Brockman is the editor of the national
bestsellers *This Idea Must Die, This Explains Everything,
This Will Make You Smarter,* and other volumes. He is
the CEO of the literary agency Brockman Inc. and lives
in New York City.

www.edge.org
@edge

BOOKS BY JOHN BROCKMAN

EDGE QUESTION SERIES

WHAT WE BELIEVE BUT CANNOT PROVE
Today's Leading Thinkers on Science in the Age of Certainty

THIS WILL CHANGE EVERYTHING
Ideas That Will Shape the Future

IS THE INTERNET CHANGING THE WAY YOU THINK?
The Net's Impact on Our Minds and Future

CULTURE
Leading Scientists Explore Societies, Art, Power, and Technology

THE MIND
Leading Scientists Explore the Brain, Memory, Personality, and Happiness

THIS WILL MAKE YOU SMARTER
New Scientific Concepts to Improve Your Thinking

THIS EXPLAINS EVERYTHING
New Scientific Concepts to Improve Your Thinking

THINKING
The New Science of Decision-Making, Problem-Solving, and Prediction

WHAT SHOULD WE BE WORRIED ABOUT?
Real Scenarios That Keep Scientists Up at Night

WHAT HAVE YOU CHANGED YOUR MIND ABOUT?
Today's Leading Minds Rethink Everything

THE UNIVERSE
Leading Scientists Explore the Origin, Mysteries, and Future of the Cosmos

WHAT ARE YOU OPTIMISTIC ABOUT?
Today's Leading Thinkers on Why Things Are Good and Getting Better

WHAT IS YOUR DANGEROUS IDEA?
Today's Leading Thinkers on the Unthinkable

THIS IDEA MUST DIE
Scientific Theories That Are Blocking Progress

WHAT TO THINK ABOUT MACHINES THAT THINK
Today's Leading Thinkers on the Age of Machine Intelligence

LIFE
The Leading Edge of Evolutionary Biology, Genetics, Anthropology, and Environmental Science

KNOW THIS
Today's Most Interesting and Important Scientific Ideas, Discoveries, and Developments

THIS IDEA IS BRILLIANT
Lost, Overlooked, and Underappreciated Scientific Concepts Everyone Should Know

THE LAST UNKNOWNS
Deep, Elegant, Profound Unanswered Questions About the Universe, the Mind, the Future of Civilization, and the Meaning of Life

HarperCollins*Publishers*

DISCOVER GREAT AUTHORS, EXCLUSIVE OFFERS, AND MORE AT HC.COM.